Contents

Page

Illustrations

Chapter 1

The Military Revolution Debate

There is one thing stronger than all the world, and that is an idea whose time has come.[1]

— Victor Hugo

It was a new theory, introduced into a profession ripe with stagnation. More than two decades passed before the broad concept beneath the theory caught fire and erupted into controversy, then debate. Today, the military revolution remains one of the most publicized, yet least understood, notions in the annals of military theory.

During an inaugural lecture at the Queen's University of Belfast in January 1955, Michael Roberts first explored the concept of revolutions in military affairs. His address, entitled "The Military Revolution 1560-1660," concerned four distinct changes in the art of war during the century-long period in early modern Europe, centering on the tactical reforms instituted by Maurice of Nassau and Gustavus Adolphus.

The gradual replacement of the lance and pike by the arrow and the musket spurred what Roberts referred to as the *revolution in tactics*. "As feudal knights fell before the firepower of massed archers or gunners," Roberts noted, a noticeable increase in the size of European armies occurred, with as much as a tenfold increase between 1500 and 1700. Eventually, as military forces expanded, "more ambitious and complex" strategies were developed to leverage the increased lethality of the larger armies; this, according to Roberts, represented a *revolution in strategy*. Finally, Roberts focused on the impact of

1

war on society: larger forces incurred a greater financial burden on the state, were exponentially more destructive in application, and required significantly increased administrative and logistics support during peacetime as well as war.[2]

Yet, what differentiated his "military revolution" from the myriad innovations of note in early modern Europe, such as formalized military education, the advent of voluminous literature on the art of war, and the postulation of scientific tenets of war? Independently, these advancements and others significantly influenced the development of military forces, yet Roberts did not view such progress as revolutionary.

Instead, Roberts categorized independent military innovations as isolated, discreet events on a developmental timeline, limited in scope and influence. Conversely, the military revolution spawned by the tactical reforms undertaken by Maurice of Nassau and Gustavus Adolphus resulted in a fundamental shift in the socio-political environment of early modern Europe.

According to historian Clifford J. Rogers, their "return to linear formations for shot-armed infantry and aggressive charges for cavalry" initiated a revolutionary chain of events that gave rise to the modern nation-state. The tactical innovations fostered by Maurice of Nassau and Gustavus Adolphus, designed to leverage advancements in technology, required greater numbers of troops with a higher degree of training and increased discipline; this in turn led to the adoption of standardized drill and uniforms. Armies soon began to adopt new, complex methods of tactical employment while at the same time growing to unprecedented proportions. As the Thirty Years' War drew to a close with the Treaty of Westphalia in 1648, large standing armies were the norm and their necessity to the state was incontrovertible.[3]

Nevertheless, it was the fundamental transformation of the character of the state that revealed the presence of revolutionary change. In order to wage war with these new-style armies, governments levied the enormous financial burden for raising and sustaining them on the populace. Never before had society been so encumbered; to manage and direct these vast resources, governments developed new administrations and significantly expanded the scope of their authority. Governments, increasingly centralized and bureaucratic, evolved to become "the paramount symbol of the modern era": the nation-state.[4]

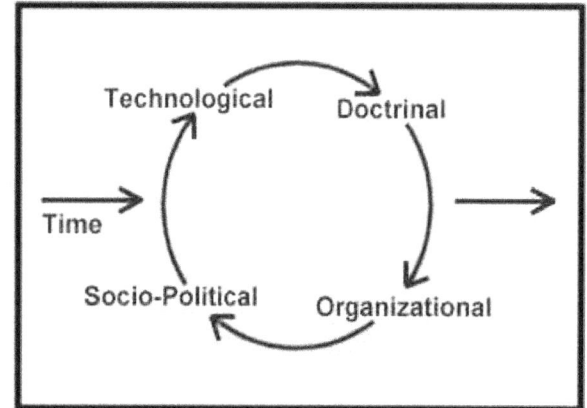

Figure 1. Michael Roberts' Military Revolution Paradigm

According to Michael Roberts' military revolution paradigm (fig. 1), the military transformation could not come full circle without the accompanying socio-political dimension. To constitute a revolution in *military affairs*, the phenomena must effect significant change in each facet of the paradigm, eventually bringing about a fundamental shift in *social and political affairs*. Clifford Rogers, in his seminal treatise on the subject, *The Military Revolution Debate*, summarized Roberts' perspective:

> At this point, Roberts' analysis of military changes merges into a consideration of their constitutional and societal impact. Larger, more permanent armies . . . "led inevitably to an increase in the authority of the state." Thus, the centrally organized, bureaucratically governed nation-state . . . ultimately grew from the tiny seed of late-sixteenth century tactical reforms. Military factors played a key, even a pre-eminent, role in shaping the modern world.[5]

Essentially, Roberts hypothesized an inexorable link between the military, social, and political dimensions reflective of the "remarkable trinity" espoused by Carl von

3

Clausewitz, the renowned eighteenth century Prussian military soldier-theorist (fig. 2). In summarizing the opening chapter of his classic *On War*, Clausewitz addressed this symbiotic relationship:

> War is more than a true chameleon that slightly adapts is characteristics to the given case. As a total phenomenon its dominant tendencies always make war a paradoxical trinity – composed of primordial violence, hatred, and enmity . . . of the play of chance and probability . . . [and] as an instrument of policy, which makes it subject to reason alone. . . The first of these three aspects mainly concerns the people; the second the commander and his army; the third the government.[6]

Roberts, while developing an illustrative model for military revolutions, reached conclusions identical to those noted by Clausewitz a century earlier. A state's military, people, and government form a complex, interdependent system; an event that affects one dimension inevitably induces change in the others, as well.[7]

For Roberts, this was the crux of his theory. His archetype of military revolution was essentially holistic in nature; to be truly considered a revolution, the effects must be realized across the entire spectrum of the model and, ultimately, would induce a *paradigm shift* in the system. Incremental or gradual change that did not affect the system as a whole would not constitute a revolution in military affairs.

For Roberts, the military revolution represented a *discontinuity event*, an abrupt and unforeseen change in the character and conduct of warfare, causally linked to social, political, and military considerations. While Roberts never specifically defined the term, he

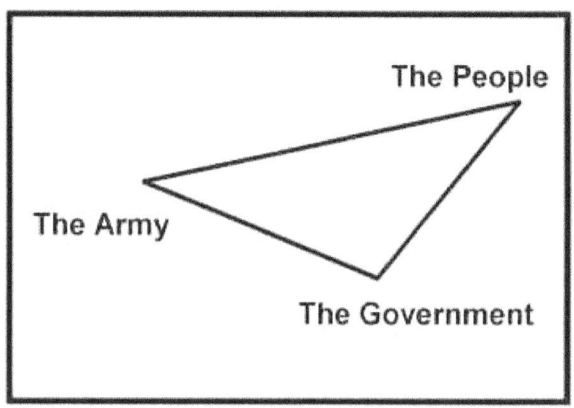

Figure 2. The Clausewitzian Trinity

4

characterized the revolution in military affairs as a sudden and fundamental departure from the customary, gradualistic accrual of military capability.

The impact of Michael Roberts' theory of military revolution was immediate and widespread. Yet, according to Geoffrey Parker, "like so many other inaugural lectures . . . [the hypothesis] would have been immediately forgotten had Sir George Clark . . . not singled out the idea for special praise as the new orthodoxy."[8] In his 1956 Wiles Lectures at Belfast, Clark proved instrumental in drawing scholarly focus upon Roberts' thesis.

Like Roberts, Clark saw a profession in decline. Military historians no longer probed the broader theoretical concepts that once defined the professional art, preferring instead to simply describe events. Rather than follow in the wake of their brethren, social historians abandoned the link between military and social history to explore other venues.

> Some few sociologists, indeed, have realized the importance of the problem; but historians tend to find their expositions a trifle opaque, and their conclusions sometimes insecurely grounded. Yet it remains true that purely military developments . . . did exert a lasting influence upon society at large.[9]

Societal associations notwithstanding, the conceptual captivation with revolutions in military affairs increased exponentially following Clark's address at Belfast. Roberts' theory "immediately found wide acceptance among early modern historians."[10] Historian Geoffrey Parker, at times a vehement critic of Roberts, hailed the concept of military revolutions as a seminal theory, a "manifesto proclaiming the originality, the importance, and the historical singularity of certain developments in the art of war in post-Renaissance Europe."[11]

Eventually, the military revolution debate spawned three distinctly separate and unique perspectives on the spectrum of change in the military continuum. Geoffrey

Parker professed the primacy of technology as the principal agent of change in revolutions in military affairs; *technologists* like Parker fostered the belief that advances in technology spur military revolutions.[12] Parker did not accept Roberts' assertion that the tactical innovations of Maurice of Nassau and Gustavus Adolphus represented a revolution in military affairs. For Parker, the advent of the *trace italienne*, or artillery fortress, was the instrument of change that drove the military revolution of early modern Europe.[13]

Roberts, conversely, acknowledged only that technology was a contributing factor in the military revolution. His perspective on the military revolution debate became characteristic of the *holistic* perspective, which fostered his adaptation of the Clausewitzian trinity paradigm as the fundamental agent of change in revolutions in military affairs. In general, holists viewed technology as an interdependent variable within a complex evolutionary system; revolutionary change was driven by a number of variables acting in concert to produce the catalyst to fire the engine of change.

Finally, the *singularists* were to be found somewhere between the widely dispersed philosophies, postulating that both technological and non-technological agents fueled the military revolution of early modern Europe. Historian Clifford Rogers, author of *The Military Revolution Debate*, postulated that a *singularity*, an isolated catalyst, served as the agent of change for the resultant military revolution. Rogers also fostered the argument that each singularity fueled a specific form of revolution, e.g., tactical, strategic, administrative, or technological, which cascaded change upon early modern Europe, combining to form a greater "military revolution."[14]

Nevertheless, of the three philosophical approaches to revolutions in military affairs, only Michael Roberts produced a paradigm that fully encompassed the complexity of the system that defined military revolutions. In fact, when templated onto Clausewitz's "paradoxical trinity,"

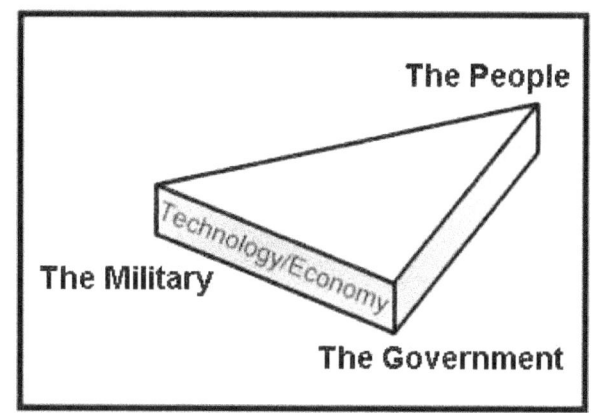

Figure 3. Author's adaptive construct for Revolutions in Military Affairs

the result is a model ideally suited to describing the fundamental interrelationship attendant during military revolutions (fig. 3). Ultimately, Roberts' holistic philosophy fashioned a descriptive representation of the military revolution phenomenon that assimilated the strength of opposing perspectives without sacrificing the integrity of his own.

In accordance with Roberts' original hypothesis, Figure 3 illustrates the inexorable link between the people, the government, and the military, all steadfastly rooted upon a firm techno-economic foundation. Any military "revolution" that excludes elements of this model is not revolutionary at all, but an isolated singularity that simply represents gradual, incremental *evolutionary* change within a dynamically complex system. To constitute a revolution in military affairs, the discontinuity must effect change across the full spectrum of the model, inducing a fundamental paradigm shift realized throughout the entirety of the system.[15]

This model for revolutionary change in the military continuum will form the basis for further analysis of historical revolutions in military affairs. Beginning with the military revolution described by Michael Roberts, the following chapter will examine past revolutions in military affairs in order to develop the context in which to evaluate the applicability of the biological phenomenon of punctuated equilibrium evolution.

Chapter 2

The History of Revolution

We therefore conclude that war does not belong in the realm of arts and sciences; rather it is part of a man's social existence.[16]

— Carl von Clausewitz
On War

In the aftermath of Michael Roberts' Queen's University lecture, historians possessed the conceptual framework within which to analyze warfare in the context of historical transformation. Jeremy Black, in his analysis of military revolutions, noted, "It offered an alternative to a narrative account [of military history], one that at once addresses the central questions of change and . . . continuity, and the causes and consequences of change."[17] For more than twenty years, Roberts' construct for revolutions in military affairs survived unscathed by criticism, a universally accepted thesis for the revolutionary impact of change.

In 1976, however, Geoffrey Parker challenged Roberts' theory with the first serious attempt at revisionist history:

> Professor Parker expressed reasoned doubts about whether those changes could be described as revolutionary, since serious inconsistencies emerge in any attempt to assess their practical impact. Why, in 1634, did the tactically conservative Spanish army wipe out the "new model" Swedish at Nördlingen? Why were the developments in tactics and strategy unable to bring the European conflict to any decisive conclusion?[18]

Had Parker exposed the Achilles' heel in Roberts' hypothesis, or were his findings consistent with the consequences expected with the revolutionary transformation of warfare? The dangerous quagmire inherent with revolutions in military affairs is obvious: the paradigm of military revolution is an analytical construct, not a panacea for failure.

Nevertheless, Parker successfully instigated a new debate concerning the interpretation of Roberts' military revolution paradigm. Historians began anew the process of examining military transformation through the ages. A broad expanse of models and definitions were in use and the result, while conducive to academic analysis, was comparatively incoherent and controversial. The further historians departed from Roberts' model, the more ambiguous and ill-defined the results became; historians and military analysts essentially portrayed any significant historical military advancement was another "revolution in military affairs."

What was lost in the process was Michael Roberts' original intent – to provide a construct for analyzing the relationship between warfare and the development of the nation-state, an appeal to both military and social historians. Roberts recognized that the military dimension could not be isolated for a reductionist examination; any change in the military had to be analyzed within a social, political, and economic context. The only legitimate model for evaluating revolutions in military affairs was one that accounted for the fragile state of equilibrium existing between the people, the military, and the government. No other theoretical construct provided that level of analytical depth and versatility.

Perhaps of greater value to the military revolution debate was the oft-forgotten fact that Michael Roberts was not primarily a military historian. Roberts, therefore, was not bound by professional tradition to consider transformation as an exclusively military phenomenon. Instead, he offered a theory that possessed a relevance to a broad expanse of the historical community. According to historian David Parrot, "Roberts' [thesis] progressed by clear stages from what appeared to be concerns of purely military significance towards issues of state-formation, national identity, centralization and bureaucracy which preoccupied a far wider spectrum of historians."[19]

Early Modern Europe

Not surprisingly, Roberts, the preeminent biographer of Gustavus Adolphus, focused his theory of military revolution on sixteenth-century reforms in tactics, strategy, and administration adopted by the Swedish king and Maurice of Nassau. These reforms, reacting in concert with subsequent political, societal, and economical changes, coalesced to fuel a revolutionary change in government and society. From humble origins on the battlefields of the Thirty Years' War, the reforms instituted by Gustavus and Maurice resulted in the birth of the most modern of political institutions: the nation-state.

In 1560, the European field of battle was dominated by massive, cumbersome infantry squares consisting of deep frontal belts of musketeers surrounding central blocks of pikemen. Such formations, often forty to sixty men in depth, overwhelmed smaller, less mobile armies; victory was as decisive as it was violent. When faced with a symmetrical foe, however, stalemate was inevitable. Standard methods of tactical employment of the period lacked the means to disrupt the "cohesion provided by this type of defense in depth."[20]

To counter traditional infantry squares such as the Spanish *tercio* or the larger blocks of the Swiss column, the reforms of Gustavus and Maurice fostered a return to shallower, linear formations reminiscent of Vegetius and Aelian.[21] In what Roberts defined as the *revolution in tactics*, the new formations enabled the infantry to fully exploit the capabilities of the musket and the pike, effectively maximizing the available firepower in a much more flexible manner.[22]

While Maurice relied upon these reforms primarily for the defense of his realm, Gustavus continued refining them for offensive employment, which he enjoyed with brilliant success. The Swede, in forbidding the use of the caracole, restored the cavalry to a role of prominence. Cavalrymen of period would engage with a singe volley of pistol fire, then wheel away from battle; Swedish cavalry revived the direct saber charge, reviving the traditional shock power of the mounted arm.

Finally, Gustavus' experimentation with gun founding resulted in the fielding of a lightweight, transportable three-pound artillery piece that would enable the Swedes to integrate close fire support with both infantry and cavalry. Gustavus' tactical reforms, designed to counter the deep formations that evolved to replace individual combat following the dark ages, produced lighter, more mobile combined arms formations capable of rapidly closing with traditional forces in decisive combat.[23] Yet, these reforms, essentially tactical in nature, were to effect profound changes throughout early modern Europe.

The reforms adopted by Maurice and Gustavus brought about a lasting effect on the training and discipline of the individual soldier. Integrating new tactics with enhanced methods of employing firepower required a renewed commitment to drill, improved fire discipline, and extensive training in the practice of combined arms.[24] Parrot summarized the consequences of the revolution in tactics:

> Elaborate, carefully taught drills were intended to speed up the rate of fire
> and to allow simultaneous volleying by multiple ranks in order both to

intimidate and to "blow open" enemy formations preparatory to a pike charge. Smaller, more numerous units could be deployed across a wide battle front in successive lines of reserves, allowing the commander far greater tactical freedom.[25]

The requisite effect on leadership necessitated increases in officer and non-commissioned officer strengths. According to Roberts, "the sergeant major of the tercio had been well content if he mastered the art of 'embattling by the square root'," while a post-reformation sergeant major had to be competent with intricate drill as well as a number of practical battle movements.[26] Drill, for the first time in modern history, became a prerequisite for success on the field of battle, and superior leadership was the foundation of that mastery.

Michael Roberts highlighted the principle of "mass subordination" as the last act in the *revolution in drill*. This tenet of the reformed army marked the successful subordination of the will of the individual to the will of the commander, a return to the collective discipline of the Legions of Rome. No longer a "brute mass" or a collection of "bellicose individuals," the armies of Maurice and Gustavus were integrated, articulated organisms with echeloned command and control structures.[27]

Reforms in tactics and drill inevitably led to changes in the strategies employed against traditional armies. Once Gustavus realized the capability to engage in, and emerge victorious from, pitched battles and engagements, strategies and methods of waging war began to change. From the time of the great Swedish king, according to David Parrot, "commanders of these newly organized armies were prepared to force an enemy into battle to achieve wider political goals."[28]

As the effects of succeeding reforms cascaded change upon these armies, commanders began to call upon increasingly large numbers of forces. States which fielded armies numbering no more than 40,000 troops in 1550 would deploy nearly one-quarter million men into battle within a matter of decades. In 1632, at the very height of the Thirty Years' War, Gustavus commanded an army of 175,000 under the Swedish flag at Nördlingen.[29]

In order to raise and maintain force structures of such magnitude, states assumed direct, centralized control of recruiting, equipping, and sustaining these great armies, while at the same time developing extensive administrative agencies to manage the increasingly complex mechanisms inherent with armies of grand stature. Armies, now garrisoned on a permanent rather than seasonal basis, placed a substantial fiscal burden on the state; those societies with a strong economic base were best suited to withstand the increased tax levies required to maintain large, standing armies and expansive governmental structures.

By 1660, Roberts postulated, the full effect military revolution had been realized. To contend with these massive, transformational armies, states developed "complicated and sophisticated systems of financial management, credit, and debt servicing" long considered a key characteristic of the modern state. "The modern art of war had come to birth, and with it the outlines of state and society that have shaped modern history."[30]

From the earliest vestiges of tactical reform, the modern nation state was born.

The Grand Armée

In the years following the Treaty of Westphalia, advancements in technology – such as the development of the flintlock musket and the bayonet – were markedly influential on the conduct of warfare, yet lacked the impetus necessary to catapult Europe into another military revolution. The ascendancy of Frederick the Great as King of Prussia marked the perfection of classic dynastic warfare, while siegecraft gradually faded into relative obscurity. Machiavelli had transformed the study of war into a social science, postulating the inexorable link between constitutional, economic, and political considerations that would one day captivate Clausewitz.[31]

In his essay concerning the transition from dynastic to national warfare, historian R. R. Palmer illustrated the unfortunate circumstances confronting seventeenth and eighteenth century dynastic states:

> The dynastic form of state set definite limits to what was possible in the constitution of armies. The king, however absolute in theory, was in fact in a disadvantageous position. Every dynastic state stood by a precarious balance between the ruling house and the aristocracy. The privileges of the nobility limited the freedom of government action. These privileges included the right not to pay certain taxes and the right . . . to monopolize the commissioned grades in the army. Governments . . . could not draw on the full material resources of their countries . . . [or] their full human resources.[32]

The dynastic army closely reflected the class structure of the state. Monarchs divided their armies into two groups: officers – motivated by duty, honor, or class consciousness – and soldiers – common men, often serfs, with long enlistments and generally considered incapable of greater service. Many armies, including those of Prussia and England, relied heavily upon the support of foreign armies. Society remained disenfranchised from the military; throughout Europe, soldiers were no more welcome in public than street beggars.

On the field of battle, the armies of Europe had once again attained a state of symmetry. With relative equilibrium at hand, large-scale, decisive battle between large standing armies became a rare occurrence. Contact with an enemy force could only be achieved with the mutual consent of both warring parties. A national army represented the political and economic power of the dynastic state and would not be committed to battle haphazardly; the cost was simply too great.

Nevertheless, all this began to change with the socio-political upheaval that shook Europe after 1789.

The loss of prestige and overseas influence that plagued France after the humiliating peace of 1763 effectively laid the seeds for both revolution and the dawn of Napoleonic warfare. Almost immediately, military thought was focused on reinvention. In introducing the principle of interchangeable parts, Jean-Baptiste de Gribeauval instituted reforms that revolutionized the use of artillery, increasing accuracy while improving mobility. Pierre de Bourcet, a staff officer in the French army and principal advisor to numerous key generals, proved instrumental in the establishment of a staff college at Grenoble.[33] François-Marie de Broglie and Etienne-François de Choiseul introduced the army division as a distinct, permanent organizational structure; the French army was subsequently reorganized from a single mass into articulated, independently maneuverable organizations.[34]

Along with technological and practical innovation came a renewed interest in the writing of military theory. Foremost among the French thinkers was Jacques-Antoine de Guibert, In 1772, at the youthful age of twenty-nine, he published his *"Essai général de tactique,"* proposing the creation of a patriot or citizen army while calling for a war of popular movement. Though his call to arms was heeded, he died in 1790 in the midst of revolution, a victim of the "reactionary, the disgruntled, and the jealous."[35]

When Louis XVI made his fateful decision to convene the Estates-General in May 1789, he unknowingly signaled a turning point in French history. In preparation for this event, Louis invited his subjects publicly express their opinions and grievances; within the unprecedented response, the liberal ideology began to take form that would propel France into revolution. What began in as a conflict between royal authority and traditional aristocratic groups evolved into a triangular struggle, with the general populace opposing both absolutism and privilege.

The formation of a national assembly from the roots of Third Estate discontent convinced the king to concede to the establishment of a constitutional monarchy.[36] Although Louis, in ordering the assembled deputies to adjourn, formally denied the existence of a national assembly, the Assembly's president, astronomer Jean-Sylvain Bailly, responded, "the assembled nation cannot receive orders." Declaring the nation alone to be sovereign, the National Assembly claimed sole authority to exercise that sovereignty.

In fact, Louis had in no way reconciled himself to such acts of revolution. But, on July 13, the partisan storming of the Bastille, the notorious former royal prison, was at once a spectacular and symbolic gesture of defiance. Elitist and commoner fought together for a shared vision, despite the traditional social chasm between them. On August 4, the National Assembly decreed an end to the feudal system representative of the *ancien régime*; on August 27, the assembly promulgated its basic principles for a new constitution in a Declaration of the Rights of Man and of the Citizen.

The constitution of 1791 had existed for less than a year when a second revolution swept across Paris with a coalition army of Prussians, Austrians, and émigrés advancing on the capitol. On August 10, 1792, partisans stormed the royal palace after defeating the garrison defending Louis. On January 21, 1793, King Louis XVI, now know simply as

"Citizen Capet," was executed in an act of immense symbolic importance. For the newly assembled National Convention, there was no turning back; from the seeds of a second revolution, a French Republic was sown.

In March 1793, the National Convention elected to expand the small, professional army fielded in 1790, calling for the increase of an additional 300,000 men. In August, the assembly decreed the *lévee en masse*, the mandatory conscription of all able-bodied, unmarried men between the ages of 18 and 25. By December 1794, more than one million men were under arms and Paris had become the largest arms-producing center in the world.[37]

In detailing the events of the period, historian Larry H. Addington, Professor of History at The Citadel, noted:

> Never before had one government commanded so much power. Never before had the revolutionary idea so kindled fires in the minds of men. The revitalized French armies carried the war . . . into Belgium, overran the Dutch Netherlands, and occupied the Rhineland. . . by 1796 only Britain and Austria among the major powers of Europe remained at war with France.[38]

France was in the midst of a military revolution, yet its significance was only gradually realized.[39] The new French army exuded a sense of passion for bravery, fighting with a "particular daring and emotion that became known as *élan*. Strategically, however, little had changed; armies remained deployed in cordon fashion, with no emphasis directed toward the concentration of forces along a single line of operations.[40]

Within the context of these tumultuous events and the War of the First Coalition, Napoleon Bonaparte rose to prominence. Austria, nearing exhaustion in 1797, finally sued for peace at *Campio Formio*, ending years of unremitting war. Only Britain refused peace with France. By 1798, a second coalition – consisting of the Ottomans, Austria, Russia and Great Britain – declared war on France, initiating a second continental war.

Returning to Paris from abroad in 1799, Bonaparte led a *coup d'etat* against the Directory, bringing the Consulate to power. As First Consul, Bonaparte became commander-in-chief of all French forces; he soon brought an end to the Wars of the French Revolution, declaring an end to hostilities with Britain at the Peace of Amiens in March 1802. In 1804, he declared himself Napoleon I, Emperor of France, combining his genius for war with the power of a supreme commander.[41]

Napoleon completed the military revolution with his greatest contribution to art of war, his ability to establish a causal link between campaign maneuver and battle. In devising a methodology with which to shape tactical advantage through campaign maneuver, Napoleon laid the foundation for *concentric maneuver*, a concept that would come to fruition with the next revolution in military affairs.

Through the application of campaign maneuver, Napoleon could force an enemy to fight on terms of his choosing, then bring the mass of his forces to bear in pitched battle. In the words of Napoleonic historian Robert Epstein, "Alexander the Great and Julius Caesar could win decisive victories, but they were unable to compel an unwilling enemy to fight at a disadvantage."[42]

Unlike Alexander and Caesar, Napoleon possessed means inconceivable to his predecessors: the army corps, formally created by the emperor on March 1, 1800. With four of these multi-division formations and an army reserve, Napoleon could conceive operations so enormous that no enemy could avoid decisive battle. By finally linking campaign maneuver and battle with the innovations of the previous century, Napoleon sparked the evolutionary process from which would emerge the operational campaign and, ultimately, the operational art of war.

Warfare, with the Grand Armée at the forefront of transformation, leaped from the shadows of the seventeenth century into the nineteenth century.

The Industrial Revolution

The great victories of Napoleon at Ulm and Austerlitz in 1805 and Jena and Auerstädt in 1806 left little doubt as to the presence of a military revolution. His victory over the Russians at Friedland in 1807, in which he drove the defeated Cossacks from the field with heavy losses, resulted in the Treaty of Tilsit, under the terms of which Russia joined Napoleon's Continental System. In the aftermath of the French Revolution, a new European political order had emerged, with Napoleon firmly entrenched in power.

Nevertheless, equilibrium was gradually reestablished as other armies began to model the reforms instituted by Napoleon. In 1812, the Emperor's disastrous campaign against Russia left him with an exhausted reservoir of men and materiel and destroyed the nucleus of his officer corps. But, of even greater import, the campaign strengthened and united his enemies against him, nations that viewed events as the first sign of weakness in the French emperor.

Disaster in Spain, an indecisive battle at Bautzen, and a defeat at Leipzig in 1813 left Napoleon's forces weakened and in disarray. With coalition forces campaigning on French soil and the fall of Paris on March 30, 1814, Napoleon abdicated and sailed for the island of Elba in exile.[43] Although Napoleon returned to France on March 1, 1815, final defeat at the Battle of Waterloo sealed his fate. The Emperor once again abdicated

20

his throne and surrendered himself to the British, who transported him to exile on the island of St. Helena in the South Atlantic, where he died six years later.

After Waterloo, the major European powers existed in peaceful coexistence for nearly forty years. The intervening period provided the opportunity for a resurgence of military thought, most notably the Prussian Carl von Clausewitz and the French Baron Antoine Henri de Jomini. Veterans of opposing armies during the Napoleonic campaigns, both men offered new perspectives on warfare. Jomini focused on the geometry of Napoleonic tactics while Clausewitz spent years brooding on the deeper socio-political implications of war. Jomini's influence was prevalent throughout the American Civil War; the deeper strategic relevance of Clausewitz required a world war to fully realize.

But the greatest engine of change grew from the fires of a different kind of revolution – a revolution based in technological advancements. The advent of steam power, advancements in mass production systems, and discoveries in metallurgy, chemistry, and physics had lasting social, political, and economic impact. Revolutionary innovations in transportation and communication accelerated the pace of national expansion and enabled near-real time transmission of information. The factory system fostered large-scale production of tools and equipment while refinements in the fabrication and manufacture of replacement components directly led to the creation of the rapid interchange of repair parts.

The Industrial Revolution is considered by many historians to be the single most significant event in modern history; without doubt, technological advancements of the period proved crucial in the development of our modern world. The Industrial

Revolution redefined the political landscape, collapsed social barriers, and shaped our economic future. Nevertheless, the military drew the greatest benefit from innovation.

The steam locomotive revolutionized land transportation, and the military potential of the rail became increasingly apparent. In time, the railhead replaced the fortress magazine as the logistics base for campaigning armies, though local forage remained the primary means of support beyond the depot. In enemy territory, the importance of seizing and securing railroads as lines of communication became a paramount concern as armies campaigned far from their base of operations, necessitating uninterrupted logistics. The steam locomotive could transport troops at a rate fifteen times faster than the foot march, conserving the energies of men and animals for battle.[44]

In similar fashion, Samuel Morse's electric telegram revolutionized communications. The telegraph evolved to meet the coordination requirements of the *commercial* rail system, providing rapid transmission of the data necessary to manage the burgeoning rail industry. According to historian James Schneider, "When armies began to move by rail, they naturally exploited the *instantaneous* command and control capabilities of the telegraph."[45]

However, the strategic advantages realized with these innovations could not be translated to tactical success. Once military forces deployed from their debarkation railheads, soldiers "still marched and draft animals still drew supply wagons" and field guns. And only the development of the Beardslee field telegraph system during the American Civil War provided field commanders with the means to coordinate the movements of their forces in areas not serviced by permanent telegraph stations.

Undoubtedly, the most significant nineteenth century tactical innovation was perfection of the expandable Minié ball in 1848.[46] When used in conjunction with percussion-cap ignition, the resulting cap-and-ball rifle rendered obsolete the flintlock, increasing range by a factor of five while virtually eliminating misfires. However, with a foreboding sense of the future, this innovation transferred the tactical advantage to the *defense*; with the addition of earthworks, defenders could pour withering fire into advancing troops formations with relative impunity. In the Civil War alone, rifle fire would account for roughly ninety percent of the casualties suffered on American battlefields.[47]

In concert, the technological innovations of the nineteenth century facilitated the evolution of the operational art. Schneider notes that this process occurred in two virtually simultaneous phases:

> The first phase was the lateral distribution of forces across a theater of operations and the emergence of a continuous front. The second phase was the deepening of the theater of operations. This led to the conduct of successive deep battles and extended maneuvers throughout the depth of the entire theater of operations.[48]

Union General Ulysses S. Grant proved to be the once Civil War commander capable of successfully conducting operational level warfare. First, on the heels of Union defeat in the Wilderness, he demonstrated the capacity for *distributed operations*, executing multiple deep maneuvers and distributed battles while advancing on Robert E. Lee's Army of Virginia. Rather than seeking positional advantage and subsequent annihilation, Grant sought freedom of action, a central tenet of the operational art.[49] Finally, in bringing the Confederacy to its knees, Grant integrated all military actions east of the Mississippi River in a series of distributed operations, effectively creating a *distributed campaign*.[50]

In perfecting the operational art, General Ulysses S. Grant unknowingly witnessed the culmination of a military revolution.

The Inter-War Revolution

The American Civil War gave way to rise of the renewed Prussian state and great German victories against the Austrians in 1866 and the French in 1870. Once again, military power gradually returned to a state of equilibrium. A humiliated French military began the process of reinvention, modeling her general staff after the Prussian example while quietly yearning for the opportunity to seek retribution.

That moment came with the firing of "the guns of August" in 1914, as the European continent erupted in full-scale war for the first time in almost a century. Despite grand plans and "proven" schemes of maneuver, the opposing forces remained in relative stasis; parallel development of firearms, magazine-fed repeating weapons, and breech-loading artillery made it virtually impossible to close with the enemy in decisive battle.[51] Neither side possessed any significant advantage in materiel, doctrine, or organization. The resultant stalemate persisted until an exhausted German state capitulated in 1918. France exacted an unforgiving toll on Germany, unwittingly initiating a vicious cycle of reinvention and retribution.

The defeat of the Central Powers and the fall of the Hapsburg Dynasty profoundly altered the political landscape of postwar Europe, leaving a power vacuum that would haunt the great powers before the close of the twentieth century. A subjugated Germany, no longer compelled to preserve the socio-political status quo, was now free to explore new methods, technologies, and theoretical concepts.

Although officially dissolved under the conditions of the Treaty of Versailles, the German General Staff survived to provide the nucleus of an elite, professional army during the postwar years. Hans von Seekt, the last chief of the general staff, preserved the personnel and training of Europe's most experienced staff by assigning sections in their entirety to other governmental agencies; von Seekt maintained a skeletal organization of sixty operations staff officers to complete the monumental task ahead.

For von Seekt, the future of warfare represented a return to maneuver-based offensive operations. Mobility was a panacea; mechanization and motorization of the German forces was a fundamental necessity. To that end, he directed the energies of the staff toward rethinking and rewriting the entire library of German army doctrine, using the lessons of the Great War as a basis for developmental thought. He directed Air Service officers to capture and assess the lessons of aerial warfare.

Ultimately, von Seekt enlisted the assistance of over 500 of the most experienced German officers "to mold their war experiences into a system of modern tactics and military organization. Although the victorious nations of World War I endeavored to redesign their own tactics as well, none assaulted the problem with so comprehensive a program as Germany. Von Seekt's decision to "retain a disproportionately high percentage of General Staff officers" in the postwar *Reichswehr* would pay dividends for the next quarter century.[52]

The collective body of doctrine produced during von Seekt's tenure stressed all facets of increased mobility, focusing on improved training methods, superior use of terrain, and constant night operations. While the French rightly concluded that victory could only be attained through the offensive, their doctrine gave little mention to maneuver concepts. Trench warfare left French tactical doctrine "frozen in time somewhere between Verdun and the autumn offensive of 1918," where it would remain well into the 1930s.[53]

The aftermath of the Bolshevik revolution in Russia also presented the newly formed Soviet Union with the awesome task of building the Red Army from the ashes of the old imperial force. Like Germany, the Soviets elected to retain former czarist officers to form the nucleus of the new army. This philosophy, remarkable as it was for a repressive state, brought together Aleksandr Svechin, V. K. Triandafilov, and Mikhail Tukhachevsky, former imperial officers whose influence would fuel Soviet thought in the coming years.[54]

The task before the Red Army staff was to develop doctrinal concepts that returned the offensive to primacy, yet within the political context that now reflected the reality of the postwar Marxist state. Soviet military strategy would consist of two elements: the political-military component, defining the purpose and character of military power, and the military-technical component, the doctrinal basis for operational methods. Not until 1927, however, did Bolshevik influence permit thought to develop beyond the realm of the political dimension.

Tukhachevsky envisioned a combined arms force of mechanized and motorized units, self-propelled artillery, and aviation to achieve breakthrough; he spurred the development of airborne forces, necessary to interdict enemy lines of communication, seize deep targets, and block the retreat of a defeated foe. Presenting the argument for successive deep operations, Triandifilov believed that decisive victory could only be attained by exploiting penetration to deliver the crucial, annihilating blow. Together, these philosophical constructs laid the foundation for the Soviet operational maneuver group and the resurgence of operational art in the twentieth century.[55]

For the victors of the Great War, effecting change was an arduous process. Despite the technological innovations wrought by a maturing industrial revolution, the realities of economic depression inevitably delayed or limited the practical application of change. Though military thought was prevalent throughout the interwar years, much of it was tainted by the euphoria of victory and misguided faith in the Versailles treaty. Truly influential thought, such as the postwar writings of J.F.C Fuller on the potential of mechanized forces, received far more respect from the Soviets and was assimilated into their evolving doctrine of operational maneuver.[56]

Ironically, perhaps, the deeper, more significant military thought of the period was born of social and political revolution. As the National Socialists rose to prominence in postwar Germany and the Bolsheviks implemented their ideal of the Marxist nation-state, the general staffs of the respective countries endeavored to fuse operational theory with emergent technology. Unfortunately for Germany, the evolution of *Blitzkrieg* theory reflected a noticeable decline in operational cognition, ultimately leading to their defeat in the Second World War.[57]

The collapse of Nazi Germany under the weight of the advancing armies of the Soviet Union and the United States marked the pinnacle of a revolution in military affairs, one characterized by a second incarnation of operational art merged with the industrial and economic might of a new world order.

The Digital Age

In the aftermath of World War II, the great powers of the world rededicated their efforts toward a lasting peace, despite the lingering angst between the western allies and the powers behind what Winston Churchill decried as the "Iron Curtain." Relative stasis existed between the great nations, both technologically and ideologically. East and West, Red and Blue.

The advent of weapons of mass destruction threatened to destabilize world order, but the Soviet development of the atomic bomb in 1949 reestablished the fragile state of postwar equilibrium. Although political, social, and military thought of the period remained transfixed on nuclear holocaust, the nations of the Warsaw Pact and the North Atlantic Treaty Organization worked collectively to limit the spread of nuclear technology while gradually easing toward détente.

The dawn of the next military revolution originated in the mind of relatively obscure electrical engineer searching for a solution to what he referred to as "the tyranny of numbers." In the last fifty years of the industrial revolution, the vacuum tube dominated technology; but vacuum tubes tended to be fragile, bulky, and power hungry, while producing considerable heat. The invention of the transistor in 1947 provided a temporary solution to part of the problem, but they "still had to be interconnected to form electronic circuits, and hand-soldering thousands of components to thousands of bits of wire was expensive and time-consuming."[58]

When other workers at the Texas Instruments Semiconductor Laboratory in Dallas, Texas left for the traditional two-week vacation period in July 1958, a frustrated Jack Kilby stayed to man the deserted facility. Working on an Army sponsored "micromodule" program to further reduce the size of electronic circuits, Kilby began experimenting with materials and processes in order to produce a semiconductor constructed of components of uniform size, shape, and material. What he presented

**Figure 4. Jack Kilby's Integrated Circuit
(Source: Texas Instruments, Inc.)**

to fellow engineers and executives on September 12, 1958 sparked a military revolution.

> What they saw was a sliver of germanium, with protruding wires, glued to a glass slide. It was a rough device, but when Kilby pressed the switch, an unending sine curve undulated across the oscilloscope screen. His invention worked — he had solved the problem.[59]

Industry greeted Jack Kilby's integrated circuit (IC) with raw skepticism. The military, however, saw the merits in his invention and sponsored further development. In 1961, the Air Force fielded the first computer to feature IC technology and the following year, the Minuteman Missile program enlisted the power of integrated circuitry.

Kilby followed his success with the invention of the first hand-held calculator, replacing the electro-mechanical desktop models of the day and soon relegating yesterday's room-sized computers to memory. The integrated circuit virtually created the modern digital industry, paving the way for a technological revolution that would literally transform every facet of society.

After a dubious flirtation with doctrine based upon the nuclear battlefield, military thought began to leverage the advent of digital capabilities, focusing on decisive battle achieved through a combination of dominant maneuver and overwhelming firepower characterized by precision weaponry.[60] Rapid advancements in technology quickly rendered obsolete the forces representative of the late industrial age. Beginning in the 1950s, the race was on to develop an insurmountable technological advantage; by the early 1980s, however, the industrial and economic might of the western powers prevailed and the Soviet Union collapsed in disarray.

The modern revolution in military affairs was in the final stages of maturity on the eve of victory during the Gulf War in 1991. In what Andrew Krepinevich referred to as the "Military Technical Revolution," the marriage of precision-guided weapons with advanced airframes such as the F-117 stealth fighter and the integration of digital fire control systems in mechanized forces produced a level of technological asymmetry unlike any in recorded history.[61] In concert with dominant maneuver theory and an unprecedented capability to deliver precise fires, the technological advantage ceded to coalition forces by an overconfident Iraqi leadership resulted in an operational victory as decisive as it was misleading.

Why misleading? According to historian Earl Tilford:

> Technology is extremely seductive and it is easy to get caught up in the exotic potential of the RMA. But in pursuit of a new way of making war, one cannot allow technological romanticism to engender visions of a mystical silver bullet which promises to sanitize war by erasing its human dimension.[62]

In the wake of the Gulf War, many believed the technological advantage demonstrated by coalition forces to be *status quo*, a panacea for future conflict. Without a contextual understanding of the nature of military revolutions, the belief in the existence of a mythical *silver bullet* existed prevailed.

The circumstances surrounding the fall of the Berlin Wall, the defeat of Iraqi forces in the Gulf War, and the resultant collapse of the Soviet Union signaled an end to another revolution in military affairs. The United States emerged at the forefront of a new world order, dominant among post-industrial age nations. However, an absolute faith in the mythical silver bullet and a moral ineptitude led America to a series of indecisive interventions during the following decade.

How might have a clearer understanding of the nature and context of military revolutions produced different results? Ironically, the answer may exist, not within the historical analysis of revolutions in military affairs, but in a *biological* theory as equally controversial as Michael Roberts' concept of military revolution.

Chapter 3

A New Paradigm

Believing as I do that man in the distant future will be a far more perfect creature than he now is, it is an intolerable thought that he and all other sentient beings are doomed to complete annihilation after such long-continued slow progress.[63]

— Charles Darwin
Life and Letters of Charles Darwin

Young Turks.

They used the term derisively, the self-acknowledged elite, spitting the second syllable as much in displeasure as disdain. How dare these *children* challenge the established status quo, the very tenets of the profession? Yet, challenge they did, and their efforts breathed new life into a science long immersed in stagnation.

Thirty years ago, the "young Turks" were a new generation of paleontologists, frustrated with a profession that had degenerated into little more than rote memorization of the fossil record and insipid, unimaginative theory. Broader conceptual and theoretical analysis had all but ceased to exist; papers of minor professional value, limited in scope and applicability, dominated meetings and conventions.

In 1971, the publication of a radical new textbook, *Principles of Paleontology*, set in motion a chain of events that would forever alter the course of contemporary paleontology. Rather than dwell on idiographic examination of fossilized remains, the text "focused on the theoretical issues of how [paleontologists] interpret the fossil record."[64]   With the subsequent publication of *Models in Paleobiology* in 1972, the "young Turks" finally came of age.

When Tom Schopf assembled and edited the myriad scientific papers for *Models in Paleobiology*, he strived to create a collaborative work that emphasized solely "new conceptual approaches to the fossil record."[65]   Nevertheless, his success was primarily attributable to a single, highly controversial article that challenged the very core of Darwinian evolutionary theory, a paper presented by two relatively unknown paleontologists, Niles Eldredge and Stephen Jay Gould.

Eldredge and Gould were characteristically representative of the generation of "young Turks."   As graduate students at the prestigious American Museum of Natural History in New York in the late 1960s, they independently studied evolutionary development in fossil invertebrates.   As their research matured, they "found that tracing evolution in their chosen organisms was difficult; most of their fossils [exhibited] no change through thousands to millions of years of strata."[66]

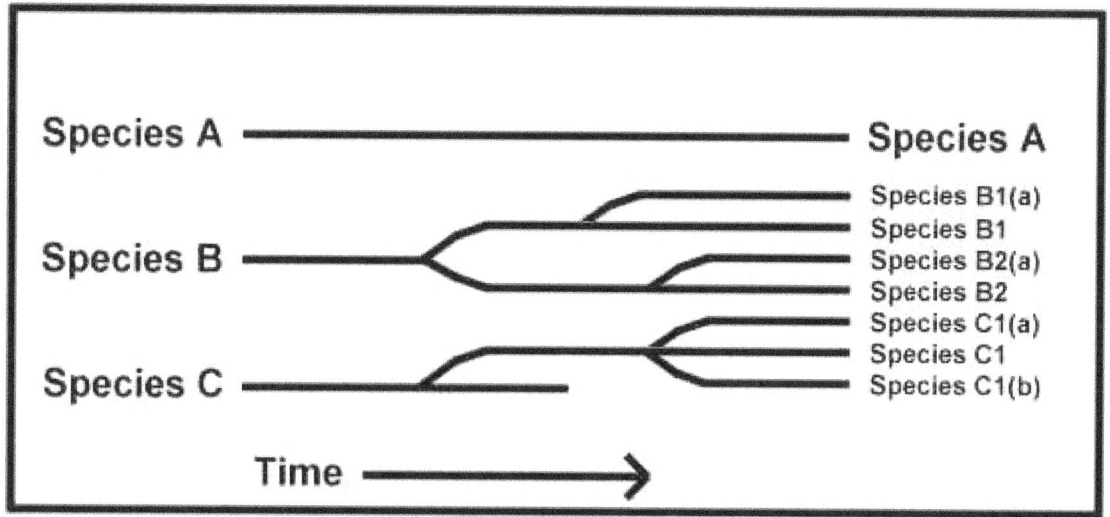

Figure 5. Phyletic Gradualism Evolution

According to Eldredge and Gould, changes within the fossil record were either insignificant or so dramatic that they warranted the definition of an entirely new species. "Most species are discreet at any moment in time. [Historical species classification] has no objective application to the evolving continua."[67] Rather than a model of gradual change within a continuum, as Darwin had proposed, Eldredge and Gould offered a controversial new paradigm: *punctuated equilibrium* (fig. 6).

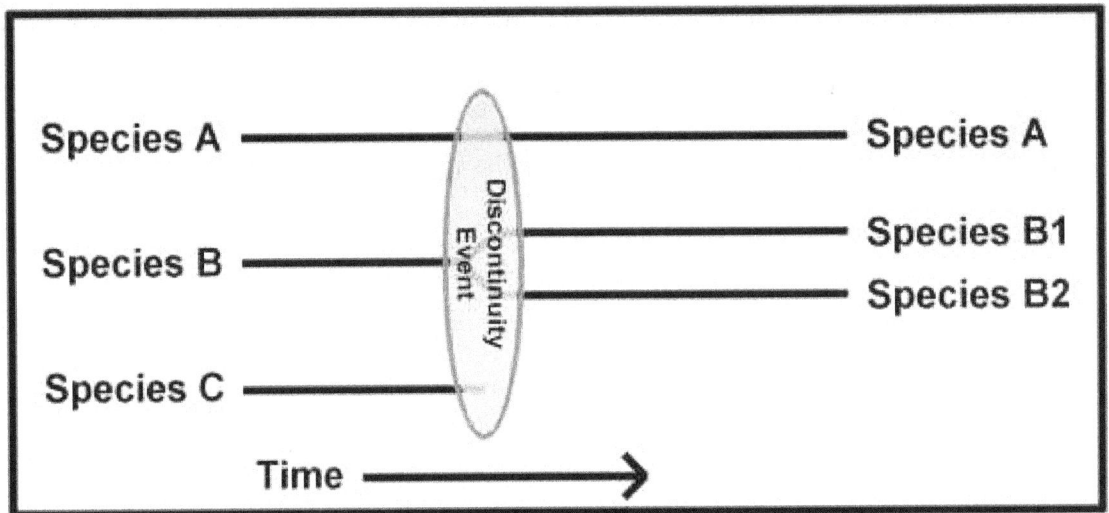

Figure 6. Punctuated Equilibrium Evolution

The essence of their research data uncoupled the fundamental tenets of paleontology, *phyletic gradualism* and *speciation*, the Darwinian model that described beneficial, gradual transitions between species in the fossil record. Also known as *sympatric speciation*, Darwin's evolutionary paradigm was the essential precept of *natural selection*. The evidence presented by Eldredge and Gould, however, suggested a model of speciation separate from adaptation.

Eldredge and Gould postulated a theory that departed from Charles Darwin's gradualistic approach to evolution, proposing that species change transpired in relatively short bursts *punctuated* by long periods of near stasis, or equilibrium. According to their theory, a species in stasis could be interrupted by sudden bursts of speciation or abrupt extinctions, but would essentially exhibit a general state of equilibrium.

Yet, how could some species linger in relative stasis for millions of years, while others died out within a few millennia? The answer, while still a matter of some debate, is a *discontinuity event*, an incident of sufficient magnitude to compel a species to adapt or suffer extinction. Such events range in scale from a rapid, unexpected climatic shift to the collision of a meteor with the surface of the planet. The scope and duration of a discontinuity event are equally unpredictable. Some species experience significant change while others remain largely unaffected; the effects of one discontinuity event may be evident within several millennia, whereas another will linger for millions of years and alter speciation across the entire spectrum of biological life.

While many scientists dismissed punctuated equilibrium as ambiguous, poorly defined, or simply irrelevant, the theory has evolved to become one of the most stimulating and provocative hypotheses in recent history. The sheer volume of literature generated from their revelation is a testament to the impact of punctuated equilibrium on the paleontological community.

Before the end of the decade, the influence of punctuated equilibrium spread to other scientific disciplines, even to fields of study totally unrelated to paleontology. Leading theorists in the fields of management, business, and finance successfully applied the fundamental aspects of punctuated equilibrium to predict and illustrate behavior of systems in their respective areas of expertise. By 1995, what originally began as a biological theory of evolution found root in the most unlikely of venues: the ongoing debate concerning revolutions in military affairs.

In his treatise on military transformation, *The Military Revolution Debate*, Clifford Rogers hypothesized the applicability of punctuated equilibrium to revolutions in military affairs. In punctuated equilibrium, Rogers believed there existed "a paradigm . . . able to provide a conceptual framework broad enough and sturdy enough to support analysis" of the complex variables inherent in military revolutions.[68] Rogers recognized what he perceived to be the probable presence of punctuated equilibrium at work during military revolutions dating back to the fourteenth century. In each instance, a historical revolution in military affairs was preceded by a relatively long period of near stasis before the sudden burst of developmental change that characterized the revolution. Rogers even noted that "a similar process of punctuated equilibrium evolution in military technology continues . . . today."[69]

Nevertheless, he chose not to explore his hypothesis further, and left the subject relatively unexamined and a matter of contemporary debate. Is punctuated equilibrium a valid paradigm for predicting and analyzing revolutions in military affairs? If Clifford Rogers was correct in his supposition, then the fundamental characteristics that define and describe punctuated equilibrium can also be templated in a similar manner with revolutions in military affairs.

Generally, punctuated equilibrium can be characterized by a system existing in a condition of relative stasis, which is interrupted by a discontinuity event that disrupts and compels a rapid change in the system. Gradually, equilibrium is reestablished within the system. Any number of subsystems may be affected; some may experience significant change while others exhibit no noticeable change whatsoever.

With an understanding of the fundamental principles that govern punctuated equilibrium, a functional paradigm begins to evolve that is remarkably similar to the phenomenon Michael Roberts described as the military revolution. Within the context of this paradigm, a comparative analysis of the characteristics of historical revolutions in military affairs will provide the construct necessary to determine the applicability of punctuated equilibrium as a predictive model for military revolution.

Chapter 4

Punctuated Revolutions in Military Affairs

I showed our article to my father. He said, "This is terrific; it will really shake things up." I replied, "Nobody will read it, and no one will pay any attention." He was right. He usually was.[70]

— Stephen Jay Gould
Living in a Punctuation

Previous analysis of punctuated equilibrium theory revealed the existence of four primary characteristics that define the evolutionary paradigm postulated by Niles Eldredge and Stephen Jay Gould. While Charles Darwin theorized that biological systems exhibit a steady state of evolution, the independent research of Eldredge and Gould clearly demonstrated that most evolutionary change occurred rapidly over relatively short periods.

According to their evolutionary model, a biological system exists in a virtual state of equilibrium until a discontinuity event interrupts the stasis of the system. While the nature and duration of discontinuity events vary, the result is the same: the discontinuity propels the system into a period of speciation, a paradigm shift in which the biological system experiences fundamental change. Following speciation, the system gradually returns to a state of equilibrium.

Why is a biological system susceptible to a discontinuity? A biological system's inherent relationship to the environment in which it exists renders the system extremely sensitive to sudden changes in the environment. Because of that fundamental

38

relationship, rapid changes in the environment catapult a biological system into a period of adaptation. The biological paradigm shift that results produces a new species adapted to the altered environment.

The central question remains: can the theory of punctuated equilibrium describe revolutions in military affairs? In Chapter 2, historical analysis of military revolutions revealed characteristics similar to those exhibited during punctuated equilibrium evolution (fig. 7).

```
┌──────────────────────────────────────────────────┐
│   Characteristics of        Characteristics of    │
│  Punctuated Equilibrium     Military Revolutions   │
│                                                    │
│  1. Equilibrium          1. Equilibrium            │
│  2. Discontinuity Event  2. Revolutionary Change   │
│  3. Speciation           3. Socio-Political Shift  │
│  4. Gradual Equilibrium  4. Gradual Equilibrium    │
│                                                    │
│  Note: Both systems will exhibit gradual change    │
│  within their respective continuum                 │
└──────────────────────────────────────────────────┘
```

Figure 7. Fundamental characteristics of Punctuated Equilibrium and Revolutions in Military Affairs

The military dimension of warfare is causally linked to its own environment, represented by the social, political, and economic dimensions within which the military exists (fig. 3). A sudden change in the nature of any one of these dimensions will fuel changes in the other dimensions, as well. A military revolution occurs when unexpected changes within the continuum of that environment catapult the military into a period of rapid change.

Prior to a military revolution, the environment described above exists in relative stasis, what is commonly referred to as symmetry. A period of revolutionary change occurs, similar to a discontinuity event, which propels the entire system into a cycle of rapid change. Ultimately, that period of change results in a fundamental socio-political paradigm shift, equivalent to speciation, followed by a gradual return to equilibrium, characterized by force modeling and shared technology.

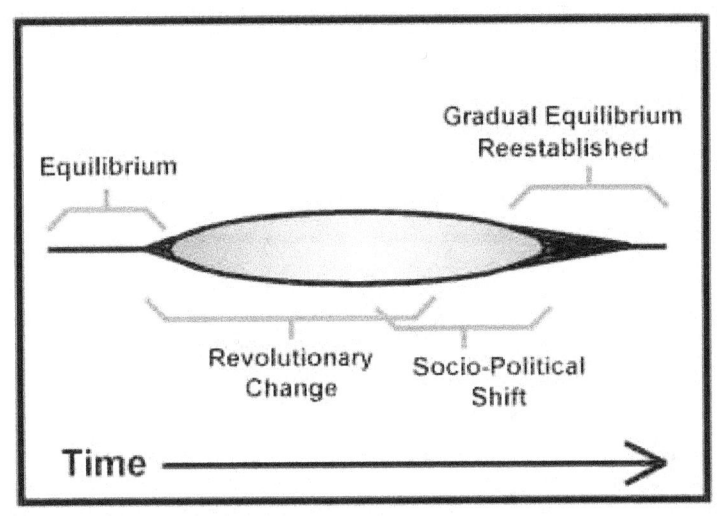

Figure 8. Military Revolution within the context of Punctuated Equilibrium theory

In the case of military revolutions, historical paradigm shifts were characterized by primarily socio-political events, such as the advent of the modern nation-state, the rise of the middle class, or a fundamentally altered balance of power. To constitute a true *revolution*, the entire system must be affected and, in turn, compel a shift in the socio-political balance of the system.

Eventually, the theory of punctuated equilibrium assimilated an element of Darwinian gradualism. Though the majority of evolutionary change occurs within the bounds of punctuated equilibrium, some speciation occurs that does not constitute punctuated equilibrium. In the same sense, change will occur within the system that bounds military revolutions; however, that change does not necessarily represent revolutionary change, but is a natural, evolutionary characteristic of the system.

In addressing the issue central to this examination of military revolution, the answer is unambiguous. Not only does the evolutionary paradigm of punctuated equilibrium provide an analytical construct with which to define the dominant characteristics of military revolutions, it also offers a predictive, interpretive model for the future.

At the same time, punctuated equilibrium offers a response to the technologists and singularists who often highlight obvious incremental change, however significant, as revolutionary. Therefore, theories such Krepinevich's Military-Technical Revolution can be considered as a phenomena separate from revolutions in military affairs. By limiting the bounds that define the scope and complexity of military revolution, the utility of the model increases exponentially.

So, what is the shape of things to come? Is a revolution in military affairs currently ongoing? What relevance does the study of military revolutions hold for the future? In response to these questions, the following chapter will address current trends in an attempt to demonstrate the versatility of the punctuated equilibrium model of military revolution.

Chapter 5

The Third Wave

It is a fallacy, due to ignorance of technical and tactical military history, to suppose that methods of warfare have not made continuous and, on the whole, fairly even progress.[71]

— Cyril Falls

With silver bullet in hand, America negotiated the last decade of the millennium by proudly launching a series of peace operations in Somalia, Rwanda, Haiti, and Bosnia, then brought the NATO coalition to bear in an effort to coerce the Serbian government to abandon its stranglehold on the former Yugoslavian province of Kosovo. In each case, results were, at best, ambiguous and, at worst, disastrous. Decisive results were consistently evasive as America grasped at straws for an elusive solution to an ill-defined problem.

For all her technological might, an industrial or even agrarian age army could incapacitate America. More so than ever before, contemporary military forces are confronted with ideologically motivated enemies entirely focused on killing them, and increasingly willing to sacrifice their own lives to do so. For the near future, that scenario defines the prevalent operational environment.

As the world enters the *information age*, America is at the forefront of what Alvin and Heide Toffler described as the Third Wave. Man entered the First Wave some ten millennia past, with the establishment of the first agrarian age society, an era when personal wealth and power were tied directly to the land. In the Second Wave, the industrial age, man's wealth and power diversified into land, labor, and capitol. In the Third Wave, the measure of man's wealth and power will be information based – the knowledge of man himself.[72]

Each wave represented a significant cultural shift for mankind, a shift that fundamentally redefined man's existence. If the world is indeed in the midst of a military revolution, then the armies that emerge will be as different from the forces of today as the great armies of World War II were from the Grand Armée of Napoleon. But the true danger for America lies in what Krepinevich described as the "dreadnought factor," the very real possibility that the United States will not be the nation to make the next crucial technological leap into the future.

Therein lies the value and relevance of understanding the nature and definition of military revolution. If the Tofflerian prediction of the Third Wave is realized in any fashion, then the power base that represents American supremacy will prove irrelevant in the future to come. The United States must focus all her efforts toward "catching the leading edge of the wave" or risk becoming a second world nation as the *information revolution* matures.

As a very real measure of American power, the military faces the same daunting task: reinvention or irrelevance. Here, the paradigm of punctuated equilibrium validates its analytical value.

In accordance with punctuated equilibrium theory, the United States is presently in the final, or gradual equilibrium, stage of the digital age military revolution. America is presently the only nation that possesses an overwhelming technological advantage;

however, the innovation gap is rapidly closing as other powers gain access to the same

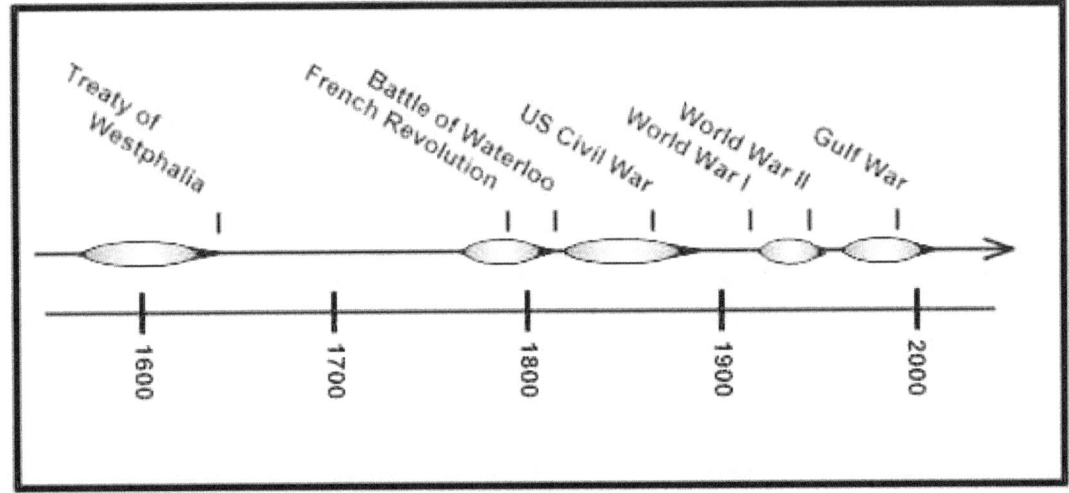

Figure 9. Punctuated Revolutions in Military Affairs

technologies. Without further reforms or technological developments, the remaining first world powers could close that gap and attain a state of equilibrium within as few as ten years. Historically, warfare conducted during periods of relative stasis is oftentimes indecisive and can trigger an unnecessary – and undesired – shift in the socio-political continuum (fig. 9). The solution to this dilemma cannot be found at the twilight of the digital age, but only on the leading edge of the Third Wave.

Directing Army transformation toward emergent or even *imagined* technology will provide the impetus to catapult America into a Tofflerian military revolution. Rather than shape doctrine to fit an unspecified or outmoded threat, military thought must now focus on preparing to fight as an *information age force*. Just as it was necessary for the "legacy force" to prepare to face a Cold War enemy, the "objective force" must be prepared for the most potent future threat.[73]

Yes, America faces an uncertain future. Nevertheless, historical analysis utilizing the paradigm of punctuated equilibrium proves that the United States can propel her efforts into a new revolution in military affairs before equilibrium is fully established in the last one. On the heels of World War II, American ingenuity provided the fuel to fire a digital age military revolution just as the flame was dying on the industrial age. That level of foresight enabled America to assume the preeminent socio-political role she presently maintains.

Today, more than ever, America stands at a crossroads with the future; a Tofflerian military revolution spans the gap between information age dominance and relative obscurity on the world's stage. And punctuated equilibrium, a paradigm for biological evolution, is the guiding light that will steer America's course. The time is nigh for the United States to accept the inevitable and ride the wave into tomorrow.

Notes

[1] Quoted from Motivational Quotes (Lombard, IL: Successories Publishing, 1984), 16.

[2] Geoffrey Parker, *The Military Revolution* (Cambridge: Cambridge University Press, 1996), 1-2.

[3] Clifford J. Rogers, ed., *The Military Revolution Debate* (Boulder, CO: Westview Press, Inc., 1995), 2.

[4] Ibid.

[5] Ibid.

[6] Carl von Clausewitz, *On War*, ed. and trans. by Michael Howard and Peter Paret (Princeton, NJ: Princeton University Press, 1976), 89.

[7] Alan D. Beyerchen, "Clausewitz, Nonlinearity, and the Unpredictability of War," in *Coping with the Bounds: Speculations in Nonlinearity in Military Affairs*, by Thomas J. Czerwinski (Washington, DC: National War College, National Defense University, 1998), 174-75.

[8] Parker, 2.

[9] Michael Roberts, "The Military Revolution, 1560-1660," in Rogers, 13. Roberts, not unlike Eldredge and Gould, viewed his profession with a sense of ire. The Clausewitzian association between the military, the people, and the state were central to Roberts' approach to military history. The departure of military historians from the theoretical analysis of the societal link to military affairs greatly concerned Roberts and was the impetus for his inaugural lecture.

[10] Rogers, 3.

[11] Parker, "The 'Military Revolution, 1560-1660' – a Myth?," in Rogers, 37.

[12] The terms used to delineate philosophical perspectives on the military revolution debate are original to the author. Such a reductionist approach to the controversy enables the author to present the arguments succinctly without necessitating a more detailed and lengthy examination not appropriate for this forum.

[13] Ibid., 38. Evolutionary improvements in the range and effectiveness of artillery fire resulted in the redesign of fortresses to withstand even the most unforgiving bombardment. Parker postulated that the technological innovation of the *trace italienne*, in combination with artillery improvements, necessitated larger forces to conduct sieges and provide security for the increasing dispersal of battlefields of the period. Ultimately, his theory produced the same results proposed by Roberts, but the primary agent of change during the military revolution was technology.

[14] Rogers postulated that the "Infantry Revolution," followed in turn by the "Artillery Revolution," the "Artillery Fortress Revolution," and a revolution in administrative affairs ultimately fueled the military revolution noted by Michael Roberts. His hypothesis led directly to a suggestion that these phenomena could be modeled using the evolutionary paradigm of punctuated equilibrium forwarded by Eldredge and Gould.

[15] The model illustrated in Figure 5 is the author's representation of the melding of Clausewitz's trinity with Michael Robert's paradigm for revolutionary change. Critical to this model is the isolation of the military within a social, political, and economic

context; no other existing process for illustrating revolutionary change accounts for these purely human variables. Both technologists and singularists failed to realize that revolutionary change within a human continuum cannot proceed without first effecting a delicate balance between the people, the government, and the military.

[16] Clausewitz, 149.

[17] Jeremy Black, "A Military Revolution? A 1660-1792 Perspective," in Rogers, 95.

[18] David A. Parrott, "Strategy and Tactics in the Thirty Years' War: 'The Military Revolution'," in Rogers, 227.

[19] David Parrot, "The Military Revolution in Early Modern Europe," *History Today* (September 1992): 22.

[20] Ibid.

[21] Roberts, in Rogers, 14.

[22] Ibid. Parrot, 22.

[23] Descriptions of the tactical reforms instituted by Maurice of Nassau and Gustavus Adolphus are relatively uniform irrespective of the source. The single greatest disparity in discussions of the period concern the fact that these formations could still face defeat under certain circumstances, as evidenced during the battle at Nördlingen in 1632. Some historians believe that a true revolution in military affairs should produce a capability unable to be defeated by forces that have not experienced transformation. Although history demonstrates that this is not the case, this theory persists today.

[24] Roberts, 15.

[25] Parrot, 23.

[26] Roberts, 14. The art of "embattling by the square root" involved the intricacies of forming large numbers of men into infantry squares for combat. Such formations, once mastered, required much less discipline and training to maintain than the smaller, mobile formations fielded by Maurice and Gustavus. As Roberts noted, deep formations required less discipline and a lower level of morale; it was surprising difficult to desert the battlefield with "fifteen ranks behind you."

[27] Ibid.

[28] Parrot, 24.

[29] Ibid.

[30] Ibid., 24-25.

[31] R. R. Palmer, "Frederick the Great, Guibert, Bülow: From Dyanstic to National War," in *Makers of Modern Strategy*, edited by Peter Paret (Princeton, NJ: Princeton University Press, 1986), 91.

[32] Ibid., 92.

[33] Robert M. Epstein, *Napoleon's Last Victory: 1809 and the Emergence of Modern War* (Lawrence, KS: University Press of Kansas, 1994), 19.

[34] Ibid., 106.

[35] Ibid.

[36] The Third Estate was one of three voting bodies in the Estates-General. Constitutional theorist Emmanuel-Joseph Abbé Sieyès declared in his treatise *"Qu'est-ce que le tiers état?"* that the Third Estate was, in fact, the oppressed French people. On June 17, the Third Estate proclaimed that they were the nucleus of a national assembly

and began to draw in deputies from the other estates. In a royal council on June 23, the king pledged to honor civil liberties, agreed to fiscal equality, and promised that the Estates-General would meet regularly in the future, although they would deliberate separately by order. France was to become a constitutional monarchy, but one in which the *ancien régime* would be conserved in its entirety.

[37] Larry H. Addington, *The Patters of War Since the Eighteenth Century* (Indianapolis, IN: University of Indiana Press, 1984), 24.

[38] Ibid.

[39] Palmer, in Paret, 112.

[40] Epstein, 26-27.

[41] Ibid., 30.

[42] Ibid., 31.

[43] Napoleon abdicated on April 11 and on April 28, set sail for Elba.

[44] Addington, 48.

[45] James J. Schneider, "Vulcan's Anvil" The American Civil War and the Emergence of Operational Art," *Theoretical Paper No. 4* (Fort Leavenworth, KS: SAMS, 1991), 52-53.

[46] French Captain Claude Minié perfected the expandable, cylindro-conoidal shaped bullet, subsequently referred to as the Minié ball, in 1848.

[47] Addington, 49.

[48] Schneider, 31.

[49] Ibid., 38.

[50] Ibid., 43.

[51] Michael Howard, "How Much Can Technology Change Warfare," in *Two Historians in Technology and War*, by Howard and John F. Guilmartin (Carlisle Barracks, PA: Strategic Studies Institute, U.S. Army War College, 1994), 3.

[52] James S. Corum, *Roots of Blitzkrieg: Hans von Seekt and German Military Reform* (Lawrence, KS: University Press of Kansas, 1992), 29-40.

[53] Ibid., 40-49.

[54] Condoleezza Rice, "The Making of Soviet Strategy," in Paret, 650-51.

[55] Ibid., 668-69.

[56] Shimon Naveh, *In Pursuit of Military Excellence: The Evolution of Operational Theory* (London: Frank Cass Publishers, 1997), 107. The influence of Fuller and B.H. Liddel-Hart cannot be overstated. Naveh notes that many historians believe them to be the impetus behind the German *Blitzkrieg* theory; Rice (in Paret, 667) argues that Fuller was translated into Russian as early as 1923, four years before the development of operational theory began in earnest.

[57] Ibid., 112.

[58] Texas Instruments, Inc, "The Chip that Jack Built Changed the World," *www.ti.com/corps/docs/kirbyctr/jackbuilt.shtml* (January 2001).

[59] Ibid. Jack Kilby was eventually recognized as a Nobel Prize Lauriat in Physics for his invention.

[60] Earl H. Tilford, *The Revolution in Military Affairs: Prospects and Cautions* (Carlisle Barracks, PA: Strategic Studies Institute, U.S. Army War College, 1995), 8-10.

Notes

[61] Ibid,, 11.

[62] Ibid., 10.

[63] Charles Darwin, *Life and Letters of Charles Darwin* (New York: Appleton, 1887), page unavailable.

[64] Donald R. Prothero, "Punctuated Equilibrium at Twenty: A Paleontological Perspective." *Skeptic* (Fall 1992): 38.

[65] Ibid.,

[66] Ibid., 39. The American Museum of Natural History in New York dominated the study of paleontology from the turn of the century and was a "proving ground" for some of the greatest minds in the field. Eldredge and Gould were examining evolutionary development in fossilized trilobites and land snail, respectively, during the formative years of their graduate program.

[67] Niles Eldredge and Stephen Jay Gould, "Punctuated Equilibria: An Alternative to Phyletic Gradualism," in *Models in Paleobiology*, edited by Thomas J. Schopf (San Francisco: W.H. Freeman, 1972), 93.

[68] Rogers, 77.

[69] Ibid.

[70] Stephen Jay Gould, "Life in a Punctuation." *Natural History* (October 1992): 10.

[71] Tilford, 14.

[72] Alvin and Heidi Toffler, *War and Anti-War: Survival at the Dawn of the 21ˢᵗ Century* (Boston: Little, Brown and Company, 1993), 15-17.

[73] Tilford, 14.

Bibliography

Books

Addington, Larry H. *The Patters of War Since the Eighteenth Century.* Indianapolis, IN: University of Indiana Press, 1984.

Alberts, David S. and Thomas J. Czerwinski, eds. *Complexity, Global Politics, and National Security.* Washington, DC: National War College, National Defense University, 1997.

Ayton, Andrew and Price, J.L., ed. *The Medieval Military Revolution.* London: Tauris Academic Studies, 1995.

Beachley, David R., Beck, Daniels C. and Troussova, Ioulia V., *Global Perspectives on the Revolution in Military Affairs: Selected Russian Views on the Changing Nature of Conflict.* Greenwood Village: Science Applications International Corporation, 1997.

Bracken, Paul and Alcala, Raoul Henri. *Whither the RMA: Two Perspectives on Tommorow's Army.* Carlisle Barracks, PA: Strategic Studies Institute, U.S. Army War College, 1994.

Chase, James and Carr, Caleb. *America Invulnerable: The Quest for Absolute Security from 1812 to Star Wars.* New York: Summit Books, 1988.

Clausewitz, Carl von. *On War.* Edited and Translated by Michael Howard and Peter Paret. Princeton, NJ: Princeton University Press, 1976.

Cooper, Jeffrey R. *Another View of the Revolution in Military Affairs.* Carlisle Barracks, PA: Strategic Studies Institute, U.S. Army War College, 1994.

Cordesman, Anthony H. and Blackwell, James. *Strategy and Technology.* Carlisle Barracks, PA: Strategic Studies Institute, U.S. Army War College, 1992.

Corum, James S. *Roots of Blitzkrieg: Hans von Seekt and German Military Reform.* Lawrence, KS: University Press of Kansas, 1992.

Creveld, Van Martin. *Technology and War*. Toronto: The Free Press, 1989.

Czerwinski, Thomas J. *Coping with the Bounds: Speculations on Nonlinearity in Military Affairs*. Washington, DC: National War College, National Defense University, 1998.

Dawkins, R. *The Selfish Gene*. New York: Oxford University Press, 1976.

DeVries, Kelly. *Medieval Military Technology*. Lewiston, NY: Broadview Press, 1992.

Eltis, David. *The Military Revolution in Sixteenth-Century Europe*. London: Tauris Academic Studies, 1995.

Epstein, Robert M. Napoleon's Last Victory: 1809 and the Emergence of Modern War. Lawrence, KS: University Press of Kansas, 1994.

Gore, John. *Chaos, Complexity, and the Military*. Washington, DC: National War College, National Defense University, 1996.

Holland, John H. *Hidden Order: How Adaptation Builds Complexity*. Reading, MA: Addison-Wesley Publishing Company, 1995.

Howard, Michael and Guilmartin, John F., Jr. *Two Historians in Technology and War*. Carlisle Barracks, PA: Strategic Studies Institute, U.S. Army War College, 1994.

Hale, J.R. *War and Society in Renaissance Europe 1450-1620*. New York: Martins Press, 1985.

Jablonsky, David. *The Owl of Minerva Flies at Twilight: Doctrinal Change and Continuity and the Revolution in Military Affairs*. Carlisle Barracks, PA: Strategic Studies Institute, U.S. Army War College, 1994.

_____. *Times Cycle and National Military Strategy: The Case for Continuity in a Time of Change*. Carlisle Barracks, PA: Strategic Studies Institute, U.S. Army War College 1995.

Kipp, Jacob W. *The Russian Military and the Revolution in Military Affairs: A Case of the Oracle of Delphi or Cassandra?* Carlisle Barracks, PA: Strategic Studies Institute, U.S. Army War College, 1995.

Kuhn, Thomas S. *The Structure of Scientific Revolutions*. Chicago: University of Chicago Press, 1970.

Lam, Lui and Vladimir Naroditsky, eds. *Modeling Complex Phenomena*. New York: Springer-Verlag, 1992.

Libicki, Martin C. *The Mesh and the Net: Speculations on Armed Conflict in a Time of Free Silicon.* Washington, DC: Institute for National Strategic Studies, National Defense University, 1994.

Mazarr, Michael. *The Revolution in Military Affairs: A Framework For Defense Planning.* Carlisle Barracks, PA: Strategic Studies Institute, U.S. Army War College, 1994.

Metz, Steven and James Kievit. *Strategy and the Revolution in Military Affairs: From Theory to Policy.* Carlisle Barracks, PA: Strategic Studies Institute, U.S. Army War College, 1995.

_____. *The Revolution in Military Affairs and Conflict Short of War.* Carlisle Barracks, PA: Strategic Studies Institute, U.S. Army War College, 1994.

Middleton, Drew. *Crossroads of Modern Warfare.* Garden City, NJ: Doubleday & Company, 1983.

Naveh, Shimon. *In Pursuit of Military Excellence: The Evolution of Operational Theory*, London: Frank Cass Publishers, 1997.

Oman, Sir Charles. *A History of the Art of War in the Sixteenth Century.* Mechanicsburg, PA: Stackpole Books, 1999; reprint, London: Greenhill Books, 1987.

Paret, Peter. Clausewitz and the State: The Man, His Theories, and His Times. Princeton, NJ: Princeton University Press, 1985.

_____, ed. *Makers of Modern Strategy: From Machiavelli to the Nuclear Age.* Princeton, NJ: Princeton University Press, 1986.

Parker, Geoffrey. *The Military Revolution.* Cambridge: Cambridge University Press, 1996.

Pillsbury, Michael, ed, *Chinese Views of Future Warfare.* Washington, DC: National Defense University Press, 1996.

Regan, Geoffrey. *The Guinness Book of Decisive Battles.* Middlesex: Guinness Publishing, 1992.

Rogers, Clifford J., ed. *The Military Revolution Debate.* Boulder, CO: Westview Press, Inc., 1995.

Rosen, Stephen P. *Societies and Military Power.* Ithaca, NY: Cornell University Press, 1996.

Schneider, James J. "Vulcan's Anvil: The American Civil War and the Emergence of Operational Art", *Theoretical Paper No. 4.* Fort Leavenworth, KS: SAMS, 1991.

Schopf, Thomas J., ed. *Models in Paleobiology.* San Francisco: W.H. Freeman, 1972.

Somit, A. and Peterson, S. A., ed. *The Dynamics of Evolution.* Ithaca, NY: Cornell University Press, 1992.

Tilford, Earl H. *The Revolution in Military Affairs: Prospects and Cautions.* Carlisle Barracks, PA: Strategic Studies Institute, U.S. Army War College, 1995.

Toffler, Alvin and Heidi. *War and Anti-War: Survival at the Dawn of the 21st Century.* Boston: Little, Brown and Company, 1993.

Troxell, John F. *Force Planning in an Era of Uncertainty: Two MRCs as a Force Sizing Framework.* Carlisle Barracks, PA: Strategic Studies Institute, U.S. Army War College, 1997.

Van Tuyll, Hubert P. *America's Strategic Future: A Blueprint for National Survival in the New Millennium.* Westport, CT: Greenwood Press, 1998.

Articles

Adams, Thomas K. "The Real Military Revolution." *Parameters* (Autumn 2000): 54-66.

Baumann, Robert F. "Historical Perspectives on Future War." *Military Review* (Mar/Apr 1997): 40-48.

Gould, Stephen Jay. "Life in a Punctuation." *Natural History* (October 1992): 10-17.

Jablonski, David. "U.S. Military Doctrine and the Revolution in Military Affairs." *Parameters* (Autumn 1994): 18-35.

Metz, Steven. "The Next Twist of the RMA." *Parameters* (Autumn 2000): 40-53.

Newman, Richard J. "After the Tank." *U.S. News and World Report*, 18 September 2000, 42-49.

Parrot, David. "The Military Revolution in Early Modern Europe." *History Today* (December 1992): 21-27.

Prothero, Donald R. "Punctuated Equilibrium at Twenty: A Paleontological Perspective." *Skeptic* (Fall 1992): 38-47.

Schneider, James J. "A New Form of Warfare." *Military Review* (Jan/Feb 2000): 56-61.

_____ and Izzo, Lawrence. "Clausewitz's Elusive Center of Gravity." *Parameters* (September 1987): 46-57.

Thompson, William R. and Rasler, Karen. "War, The Military Revolution Controversy, and Army Expansion." *Comparative Political Studies* (February 1999): 3-32.

Internet Articles

Texas Instruments, Inc. "The Chip that Jack Built Changed the World." *www.ti.com/corps/docs/kirbyctr/jackbuilt.shtml* (January 2001).